STARTING OUT

I0794401

baby TORTOISES

MEG GREVE

CREATIVE EDUCATION • CREATIVE PAPERBACKS

CONT

ENTS

I AM A HATCHLING.

I am a baby tortoise (TOR-tuhs).

Hard, bony scutes cover my shell.

I hatched from an egg. I used a special tooth to break out.

My mom dug a nest for her eggs. But I do not stay there. I will take care of myself.

During the day, I look for plants to eat.

I sleep at night.

I stay on land.
I do not swim.

My shell is not flat like a turtle's. It is round.

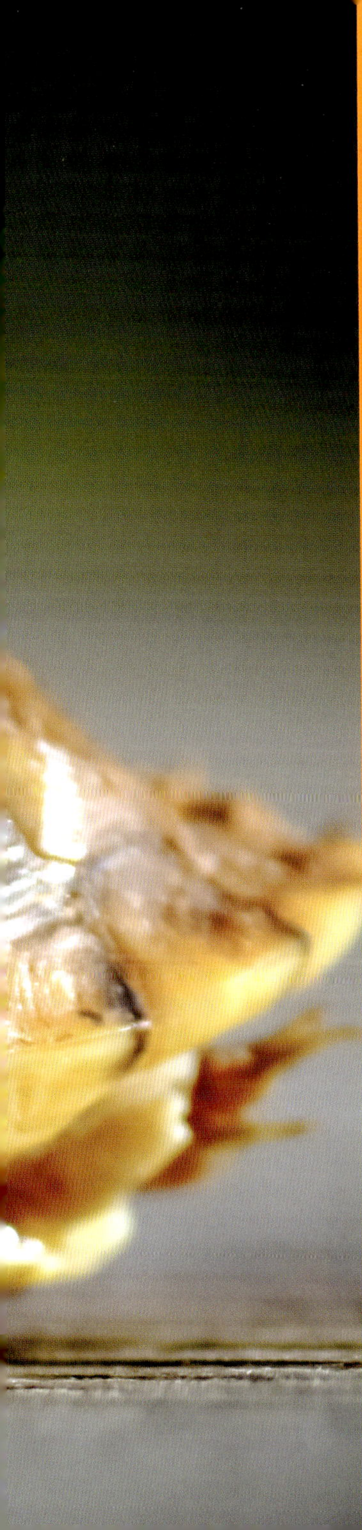

I hide inside my shell.
It keeps me safe.

I may live for 100 years or more!

SPEAK AND LISTEN

EE-EE!

Can you speak like a hatchling?

Baby tortoises squeak and chirp.

Listen to these sounds:

https://www.youtube.com/watch?v=klgVo6h5fng

Now it is your turn!

TORTOISE WORDS

hatched: broke out of an egg to be born

nest: a hole dug in the dirt by a mother tortoise to keep her eggs safe

scutes: hard, bony plates or scales that cover and protect a tortoise

tortoise: a turtle-like reptile that lives on land

READING CORNER

Kellett, Jenny. *The Ultimate Turtles Book: Discover the Shelled World of Turtles and Tortoises.* Melbourne, Australia: Bellanova Books, 2023.

Riggs, Kate. *Tortoises (Amazing Animals)*. Mankato, Minn.: Creative Paperbacks, 2022.

Somaweera, Ruchira, and Stephanie Warren Drimmer. *The Ultimate Book of Reptiles.* Washington, D.C.: National Geographic Kids, 2023.

INDEX

PUBLISHED BY CREATIVE EDUCATION AND CREATIVE PAPERBACKS
P.O. Box 227, Mankato, Minnesota 56002
Creative Education and Creative Paperbacks are imprints of The Creative Company
www.thecreativecompany.us

LIBRARY OF CONGRESS CATALOGING-IN-PUBLICATION DATA
Names: Greve, Meg, author.
Title: Baby tortoises / Meg Greve.
Description: Mankato, Minnesota : Creative Education and Creative Paperbacks, [2025] | Series: Starting out | Includes bibliographical references and index. | Audience: Ages 4-7 | Audience: Grades K-1 | Summary: "Introduce beginning readers to the hard-shelled world of baby tortoises with this life science starter. Includes photos, a labeled animal diagram, "Make a Noise" section, glossary, and further resources"-- Provided by publisher.
Identifiers: LCCN 2024014671 (print) | LCCN 2024014672 (ebook) | ISBN 9798889892113 (library binding) | ISBN 9781682775776 (paperback) | ISBN 9798889893226 (ebook)
Subjects: LCSH: Testudinidae--Juvenile literature. | Testudinidae--Infancy--Juvenile literature. Classification: LCC QL666.C584 G74 2025 (print) | LCC QL666.C584 (ebook) | DDC 597.92/4--dc23/eng/20240417
LC record available at https://lccn.loc.gov/2024014671
LC ebook record available at https://lccn.loc.gov/2024014672

DESIGN AND PRODUCTION
Design by Rhea Magaro
Production by Beeline Media and Design, Inc.
Art direction by Tom Morgan

PHOTOGRAPHS by Alamy Stock Photo/Jeffrey Schwartz, 10, Lindsey McDonnell, 8; Shutterstock/Bohbeh, 13, Eric Isselee, 11, 14, fivespots, 2, NPatricia, 9, seasoning_17, cover, 4, 5, 6, 7, 12

Printed in China